Paul Middleton first encountered the Middle East in 1984 and 1986, working on moshavs throughout Israel, and also with Palestinian money changers in East Jerusalem. A few years later, while living in Egypt, he travelled extensively in that country as well as making several visits to the West Bank. He spent the following two years living in Yemen, studying Arabic, before returning to complete an honours degree in Arabic and Middle Eastern and Islamic studies at the University of Durham in England. More recently, he spent over five years working first in Saudi Arabia and subsequently in Dubai in the United Arab Emirates. Paul lives in the north of England with his wife and three daughters.

A Brief Guide to
The End of Oil

Paul Middleton

ROBINSON
London

Constable & Robinson Ltd
3 The Lanchesters
162 Fulham Palace Road
London W6 9ER
www.constablerobinson.com

First published in the UK by Robinson,
an imprint of Constable & Robinson Ltd, 2007

A copy of the British Library Cataloguing in Publication Data
is available from the British Library

ISBN 978-1-84529-659-9

Printed and bound in the European Union

1 3 5 7 9 10 8 6 4 2

Contents

Introduction

Like almost everyone else on the planet I never used to give oil a second thought. At least not in the way it affected my life, my very comfortable Western life. Hey, Gulf Wars were inevitable and it was a terrible shame about the oil well fires and all that, but it didn't affect my life, really (apart from some spikes in petrol prices, which bugged me). Besides "we" gave those Iraqis a good seeing to and returned the region and the oil to its previous status quo; "they" supplied "us" with oil, and we used it to have a great time.

And that's about it. As long as I'm OK right now, I don't much care; and I'd say that's about right for 99 per cent of people living in the West.

But the world is changing, and issues like global warming and the "War on Terror" (silly though I think the term is) have brought our use of energy – and therefore oil – to the attention of many, if it hasn't yet spurred much action.

We're starting to get the feeling that something's going wrong, that something's happening that we can't quite put our finger on – perhaps because we don't want to – and that at some point in the future we might have to change the way we live. And that is not something any of us want to do without an awful lot of good reasons; and then some.

And that's precisely what this book aims to do. To take a no-nonsense look at oil; where it comes from, who's got it, how much is left, who uses it all, and why it is damaging our planet, if indeed it is.

So *The End of Oil* is a little book of reasons. Not reasons to stop doing anything – not yet, at least – but reasons to change the way we view our use of oil.

Chapter One

The Industrial Revolution: Why is oil such a big deal?

Back in the days when we foraged and collected fire wood for our homes things were very different; so different, in fact, that our lives then bore very little resemblance to our lives today. Indeed we probably couldn't live like that again, even if we wanted to.

That's because in the days before the industrialization of the world we were well on our way to burning every tree in the land, just to keep warm and to cook our food.

I blame the Romans, really, they were the ones responsible for the deforestation of Britain. Yes, we had a splendid "rain forest" thing going on long before - well, at the same time - as the folks in the Amazon. You could hardly move for trees in England.

This bountiful supply and the fact that we could harness it to create heat could mean only one thing: we would keep chopping down trees. But not just for cooking and heating. We love making stuff, too: furniture for our homes, plates, pots, pans, clothes (assorted colours of course), shoes, etc. - you get the picture. All of this stuff required some degree of heat to bend, bake, harden, and dye. And as our population grew from just under two million (1.9 million according to the Domesday Book of 1086) to 5.74 million in the 1750s it put serious pressure on our limited resources. Indeed, as Richard Heinberg points out in his book *The Party's Over*, "their successes [the Europeans] . . . meant that Europe, by the sixteenth century, was comparatively crowded and resource depleted.'[1]

As the trees started to run out we found and burned other stuff, coal mainly. And this was our move from utilizing the living natural resources of the planet to those that were long dead; the time when we started to burn fossil fuels: those naturally occurring carbon-containing materials, which, when burned, produce heat (directly) and energy (indirectly). Fossil fuels can be classified according to their respective forms: solid fuels (coals); liquid fuels (petroleum, heavy oils, bitumens); and gaseous fuels (natural gas). These fossil fuels have been created from decomposed vegetation which has been under great pressure for hundreds of millions of years, since well before the dinosaurs roamed, hence the name, "fossil" fuel.

And the rule of thumb, when it comes to coal, is that the older the coal and the more pressure it's been under, the hotter it will burn. We didn't quite grasp that initially, however, so we began by burning the peat-line coal, the stuff lying closest to the surface, which was easiest to get to. It didn't burn very hot (though it did burn hotter than wood) and it created pungent smoke and soot. We did catch on eventually that the best coal was usually a few hundred feet under the ground, which presented a few engineering challenges. Even though digging was something that the great English Northerner was very good at, there was no getting away from the fact that the deeper the mine - and the better the coal - the more prone it was to flooding. Our solution was to build a steam-powered water pump, something that could produce enough mechanical power to suck gallons of water out of a mine hundreds of feet below ground around the clock. This was achieved by Samuel Newcomen in 1708 with a crude, steam-powered water pump. Problem solved: dry mines, loads of high quality coal, and a very rich Samuel Newcomen.

The glorious Newcomen pump not only gave us abundant coal, it also ushered in the start of a whole new world of machines, the true beginning of the industrial revolution; the moment when machines began to replace the direct physical efforts of people. Coal was the fuel that powered the industrial revolution, a revolution that changed the social, economic, and political landscape.

And what did we do with this new found energy? We produced more stuff, of course: more food, clothes, building materials, carriages, you name it, we made it. And, as our relative wealth increased, we produced more people, loads more people. From the turn of the nineteenth century the population of England increased dramatically, from ten million in 1801 to 30 million in 1901 and 50 million by 2001.

A simple and compelling pattern emerged: the more energy we produced, the more stuff we made, and the more stuff we made, the more we wanted. And the more we wanted, the more energy we used to make more stuff.

Lovely, a virtuous circle of consumerism. Or, more correctly, we developed a dependency on coal and the energy it produced. We loved our new world of energy and it led to an explosion in mass transport like steam trains[2] (now even more efficient with hotter-burning coal) and the development of a mass transport system to replace the rather slow canals, a system that allowed us to travel around the country at speed and in comfort and which also allowed goods to be transported. The steam revolution spread to the high seas, too, and Britain's very substantial merchant navy and its Royal Navy which had been predominantly powered by sail, was soon brought up to date with marvellous new steamers, among them the Royal Navy's new Royal Sovereign Class steamers.[3]

And then, of course, there were those other great inventions of the time, like the electric light bulb (1880) and the telephone, both of which, naturally, required an electricity supply.

Indeed, on the back of his electric light bulb enterprise,[4] Thomas Edison formed the Edison Electric Illuminating Company to build electricity generating plants in New York City. The first central power plant, on Pearl Street Station in Lower Manhattan, began generating electricity on 4 September 1882.[5]

So the coal mines were busy, very busy. At the

beginning of the eighteenth century the average Englishman was responsible for the use of less than half a ton of coal a year; by the turn of the twentieth century he would be using more than four tons a year – an eight-fold increase. Given the sheer abundance and cheapness of coal it became "the fuel of choice". The smog was getting to be a bit of a pain, but we had trains and telephones!

The point here is that coal was adopted as a new energy source for three primary reasons: the first was that wood was running out and therefore becoming increasingly expensive – the preserve of the rich; two, coal was abundant (thanks in part to new mining technology) and therefore cheap; and three, it burned hotter than wood (which meant that it was an "enabler" in the industrial revolution).

It just so happened that coal had a higher energy density, but it wasn't planned that way. It was a process of discovery and invention, of change. And, like most change, the transition to coal was pretty tough for the average Englishman, as Paul Roberts points out in his book *The End of Oil* – "sixteenth century Englishmen nearly revolted at having to burn sooty coal instead of wood."[6]

Moving from one energy source to another is a difficult process, even when the socio-economic environment is relatively simple and the new energy regime is "better", i.e., it yields more energy. It also makes the point that the driving force behind change was, and remains, our desire for cheap energy so that we can make loads of great stuff.

And, this in turn led to our increasing dependency on cheap energy, and our desire to control the source of that energy. Governments were only too aware of their growing reliance on fossil fuels and determined that their respective countries should dominate their sources of energy, for energy equalled power.

The advent of the industrial revolution added great impetus to the already booming British Empire, which was, after all, an exercise in power and control, aimed at spreading British influence around the globe, something that simply would not have been possible without unfettered access to the world's energy resources. By the turn of the twentieth century, competing national and imperial interests had fuelled an arms race in Europe, which, in turn, compelled Britain to turn to a more powerful source of energy.

That source of energy was oil, which was to coal what coal was to wood. It burned hotter and cleaner than coal - not only that, but it also squirted out of the ground under its own pressure. No mining, no flooding of mines, no trouble at all; just line up the barrels and away you go. Oil wasn't something new, of course. Many civilizations had used oil - the oil that found its way to the surface of its own volition - for pouring on their enemies or use as fuel. The difference now was that the technology existed to extract it - by drilling - to refine it, and to consume it on an industrial scale.

The first commercial oil well was drilled in north-western Pennsylvania in 1859. Known as the Drake Well, after "Colonel" Edwin Drake, the man responsible for the well, it was drilled for the sole purpose of finding oil for refining into lamp fuel. But probably the most famous of the "pioneering" oil wells was the big find in Texas at a place called Spindletop. At the time of its discovery coal was still "king", its abundance and cheapness seeming to make it irreplaceable as an energy source. That all changed in Texas on 10 January 1901. Spindletop was soon producing 5,000 barrels of oil an hour; up until that time an oil well had been considered "good" if it produced fifty to a hundred barrels a day!

The USA quickly became the largest oil producer in the world and normal, rational men started putting bull horns on the bonnets of their cars, smoking unfeasibly large cigars, and wearing stirrups to the office.

Around the same time oil was being discovered in huge quantities in the Middle East. The Anglo-Persian Oil Company (APOC) was founded by Britain following the discovery of a large field in Persia. Britain saw the light and converted the Royal Navy from coal to oil power.

And that was just the start. What about cars? Though the very first self-powered road vehicles were powered by steam engines,[7] Gottlieb Daimler invented what is recognized as the prototype of the modern petrol engine - with a vertical cylinder, and with petrol injected through a carburettor (patented in 1887). Daimler first built a two-wheeled vehicle, the *Reitwagen* (Riding Carriage) with this engine and, a year later, he built the world's first four-wheeled motor vehicle. The car had been born. By 1913, there were one million cars and trucks on the roads of the USA and Europe. By 1927, the Ford Motor Company alone had built more than 15 million Model T cars.

Oil transformed our lives and the way we use energy. As Richard Heinberg puts it, "If we were to add together the power of all the fuel-fed machines that we rely on to light and heat our homes, transport us, and otherwise keep us in the style to which we have become accustomed, and then compare that total with the amount of power that can be generated by the human body, we would find that each American has the equivalent of over 150 'energy slaves' working for us twenty-four hours each day. In energy terms, each middle class American is living a lifestyle so lavish as to make nearly any sultan or potentate in history swoon with envy."[8]

By 1929, demand for oil was running at four million barrels a day.

Notes

1. p. 21.
2. On 21 February 1804, Richard Trevithick's pioneering engine hauled ten tons of iron and seventy men nearly ten miles from Penydarren, South Wales, at a speed of five miles per hour.
3. The ships of the Royal Sovereign Class were built under the Naval Defence Act of 1889, which provided £21 million for a vast expansion program.
4. Joseph Swan, a British scientist, demonstrated the first electric light with a carbon filament lamp. A few months later, Thomas Edison made the same discovery in America.
5. Pearl Street had one generator and it produced power for 800 electric light bulbs. Within fourteen months, Pearl Street Station had 508 subscribers, and 12,732 bulbs.
6. p. 9.
7. In terms of that definition Nicolas Joseph Cugnot of France had built the first "automobile" in 1769.
8. *The Party's Over*, Richard Heinberg, p. 31.

Chapter Two

Addicted: How much oil do we use and where does it all come from?

Predictably, we use loads of oil. Global consumption is running at about 84 million barrels per day (mbpd) and growing at over 2 per cent per annum. So, a billion[1] barrels lasts us just twelve days.

And who are the big hitters, the hungriest? Well, not surprisingly it is the USA, topping the charts at over 21 mbpd.

Rank	Country	mbpd	%
1.	United States	21	25
2.	China	6.4	8
3.	Japan	5.6	7
4.	Russia	2.8	3.5
5.	Germany	2.7	3.4
6.	India	2.3	2.9
7.	Canada	2.2	2.8
8.	South Korea	2.2	2.8
9.	Brazil	2.1	2.6
10.	France	2.0	2.5
11.	Italy	1.9	2.4
12.	Saudi Arabia	1.8	2.3

How much we use obviously depends very much on who we are and where we live. The average person in the USA, for example, uses 7,500 gallons of oil per year while the average for a person in China is just 800 gallons per year. Naturally, this has everything to do with population and the relative penetration and usage of things like cars, etc.

In China in 1995, for example, cars accounted for just 5 per cent of personal transport while bicycles still accounted for 33 per cent. Great for keeping overall consumption down but we'd all be living in cloud cuckoo land if we didn't believe that almost every Chinese person would like a nice new, shiny car. Indeed by 2000, cars accounted for 15 per cent of personal transport, and there are projected to be 200 million cars on the roads of China by 2020.

At present, China, with 20 per cent of the world's population only consumes 6 per cent of the oil, while the USA with just 5 per cent of the world's population consumes 25 per cent.

The point is that the demand for oil is going to grow, not decline, and the driver will be development, as it has always been.

The use of energy from oil in the USA

Cars	48%
Trucks	13%
Plastics etc.	10%
Aircraft	7%
Heating	5%
Other uses	17%

Of the oil the USA burns every day, half is used to power cars – the good old commute. That seems like an awful lot, particularly when compared with flying. But that's the reality, we love our cars. We just love them – I know I do. Who wants to stand (or sit, if you're lucky) on a crowded train or bus when you can enjoy the comfort, privacy, and entertainment of your own wheels, even if it means sitting in traffic? Almost no one. Just look out of the window and do some calculations of your own; the only reason most people use public transport is because they can't afford a car.

As Greg Greene notes in his documentary *The End of Suburbia,* "the whole post-war American way of life ended up being centred on the suburb; the nice house with a nice lawn and the picket fence; a nice car in the drive . . . it stopped being about escaping tyranny or finding democracy. Instead, it became about achieving a way of life that was propped up by cheap energy."

So millions of cars have been – and continue to be – built. Mr Ford is famous for having made the car a necessity, rather than a luxury. Everyone simply has to have a car. And more cars mean more roads (freeways, even) and more places to live and work. People don't have to live in cities any more to get a job; they can commute. Fantastic! We can live in the country and work in a city. What could be better? Millions of homes were built in suburbia for millions of workers who commute to work each day. And that's pretty much the state of play today. Most of us do just that: commute, which would be practically impossible for most people without a car.

Cheap as chips

As we know, the age of oil really got going in Texas. And while it may be true that Texans like to wear cowboy clothes, few of them are still cowboys; rather, they are serious oil men and women. So, with lots of oil and the help of some committed entrepreneurs, the USA developed not only the expertise for extracting, refining, and distributing oil products, but also the first truly modern, highly mechanized nation state. Other countries were doing it, too, of course, particularly in Europe, but the Americans led the way with great drive (no pun intended) and huge investment. They mass-produced cars that the average Joe could afford, they built highways and cities around the car (the classic grid system), they built commuter belts, and, by the

1950s, the USA had built a super-economy.

But for a while the oil business in the USA struggled. At the turn of the nineteenth century, the USA was building a new, free market economy that gave rise to mass production and consumption - a model that would be replicated around the world. And as the US economy grew, so did demand for commodities like petrol. The USA developed the technology for oil exploration and extraction faster than anyone else because, firstly, the country had seemingly inexhaustible reserves of oil, and, secondly, it was an enormous country, ripe for development, and energy was key to this.

A messy business

At the end of the nineteenth century the oil business was split between producers (the actual well owners who put the stuff in barrels), the refiners (who turned the oil into something customers could use, like petrol), and retailers (people who ran petrol stations and the like). As you might imagine, this wasn't the most efficient way of getting oil to market and it led to wild price fluctuations and supply problems. Having noted this, a smart and ruthless businessman by the name of John D. Rockefeller calculated that if one owned the well, the refinery, *and* the petrol station one would be able to manage supply much more effectively and in that way control the price. It worked a treat, as Paul Roberts points out in *The End of Oil*: Rockefeller "perfected the now standard strategy of being the lowest-cost producer, making his profits through ever larger sales volumes, while mercilessly, and often illegally, undercutting his competitors." Doesn't sound like an oil company, does it?

Oil production and consumption of the top twenty nations by production, as a percentage of global production/consumption

Production	Consumption	
Oman	1.3%	0.1%
Libya	1.9%	0.3%
Indonesia	1.9%	1.4%
Algeria	2.0%	0.3%
Brazil	2.1%	2.9%
Iraq	2.9%	0.6%
Nigeria	3.0%	0.4%
Kuwait	3.0%	0.4%
United Kingdom	3.4%	2.3%
United Arab Emirates	3.4%	2.3%
Canada	3.6%	2.2%
Venezuela	4.1%	0.7%
European Union	4.3%	19.1%
China	4.4%	6.0%
Norway	5.5%	0.2%
Mexico	4.8%	2.0%
Iran	5.0%	1.7%
Russia	9.7%	3.4%
United States	10.7%	25.9%
Saudi Arabia	11.6%	1.9%

Source: *CIA World Factbook*

By 1914 Standard Oil owned 90 per cent of the US market, and a great deal of the international market too, and the US government decided that Rockefeller's Standard Oil Company was a monopoly and forced him to split it, creating companies such as Exxon and Mobil in the process. This "vertical integration" remains the preferred model today, though, as we shall see later, non-US producers did tire of it.

This development, centred on the USA, achieved one important thing for the average oil consumer: it created the unwavering belief that oil was both cheap and abundant. It was on this belief that we built our future.

A black ocean

So where is all this oil and how much does each country produce?

Oil is everywhere. Even though its existence depends on very specific geological conditions[2], oil is to be found in a surprising number of different places.

The chart shows just how diverse the range of producer countries is, and, of course, the pronounced disparity in some cases between how much some countries or regions produce and how much they consume.

The best known oil fields are, of course, in the Middle East (Saudi Arabia, Kuwait, Iraq, etc.). Others are to be found in Nigeria, Venezuela, Norway, the USA, Russia, China, Chad, and so on. The differences between these oil fields lie in the amount of oil they produce, the quality of the oil, and how easy it is to extract (from the ground or the seabed) and deliver to refineries.

Each variety of oil has a name: US West Texas Intermediate, North Sea Brent Blend, Algerian Saharan Blend, Indonesian Minas, Nigerian Bonny Light, Saudi Arabian Arab Light, Fateh from Dubai, Venezuelan Tia Juana Light, Mexican Isthmus, and so on.

And the quality of a particular oil determines its cost.

West Texas Intermediate, for example, is generally priced at about $1 to $2 more per barrel than Brent Blend (the general rule of thumb is the "lighter" and "sweeter" the oil, the more valuable it is).[3]

If the oil is easy to extract, better still. Texas oil is sweet and light and, at least when it was first extracted, simply "gushed" out of the ground. Perfect! Great oil that costs virtually nothing to extract. This is the case, too, for Saudi Arabia's Arab Light where a barrel of oil costs less than $2 to produce; Iraq is the only country where the cost of production is lower.

That's cheap, and even with appealing margins for refiners and retailers, it still means cheap petrol for everyone.

So, until the 1970s, oil that wasn't good quality and didn't simply gush out of the ground wasn't worth extracting as there was no profit in it. That was, until oil prices started to rise above their traditional price point of about $5 a barrel.

Food and oil

Food production in the twentieth century has increased dramatically, in line with population growth. Intensive farming methods have been adopted to get more food yield from every acre of agricultural land. In fact, the rate of growth of agricultural productivity has been increasing at about the same rate as oil consumption, about two per cent a year. We now produce three times more food than a hundred years ago.

Producing food is now an industrial process, making use of diesel-powered tractors to plough, fertilize, sow, spray, and harvest crops. And the nitrogen fertilizers that are used are made from gas, and the pesticides from oil.

The trucks that transport everything from seeds to market-ready produce run on diesel, and the food sold is supermarkets is, more often than not, wrapped in oil-derived plastic packaging. We then drive to supermarkets in our cars to buy our food. It's one big oil- and gas-powered market.

In the natural world the food equation is a simple one. The energy expended in producing food is slightly less than the energy received in consuming it; life has existed according to this positive energy equation for millennia.

Not any more, however. The energy now required to produce most of our food means we're firmly in the red: expending far greater energy in producing food than we could possibly recoup by consuming that food. This negative energy equation is only possible because of access to cheap fossil fuels.

The effects of this energy-intensive food production have also changed the face of farming across the globe, forcing farmers to work larger and larger areas of land to achieve any kind of economy of scale. This pattern has been replicated around the world as demand in rich nations for agricultural produce from the developing world has increased. This demand gave rise to the "Banana Republics" in the early part of the twentieth century when Caribbean and Latin American countries were bribed, threatened, and cajoled to turn their rural smallholdings into vast plantations to grow bananas for export. This, of course, had devastating effects on the local peasant populations and reinforced the centralization of power in many of the countries concerned.

The same thing happened with regard to the

production of tea, coffee, and sugar in Asia. Energy-intensive food production also explains why, in the UK, butter from New Zealand, half way around the world from the UK, is cheaper than locally produced butter

Farming with livestock is even more energy-intensive than growing crops, as is the fishing industry. Indeed, we have become so efficient at fishing with our super-trawlers that fish stocks are in serious decline throughout the world. The long-term effects of our fossil fuel-driven food supply are already being felt in declining yields and surpluses in producer countries. There is always a point of diminishing returns, and, in food production, this has been reached, an alarming situtation when one considers the demands that the growing population of China alone will put on food supply in the next decade.

Add to that a significant rise in oil prices and the scenario becomes even more complex. Our ability to feed ourselves is currently inextricably intertwined with our use of oil; declining oil reserves will impact on world food production in ways which it is difficult to predict; the only certainty is that great changes lie ahead.

Notes

1. US billion (one thousand million).
2. The correct temperature, neither too hot nor too cold, is required to "cook" the organic material that makes oil, and the right kind of rock is also required – porous to allow the oil to flow upwards, with a suitable cap-rock to hold it in place, and so on.
3. An oil's API gravity is a measure of how heavy or light a petroleum liquid is compared to water. If the figure is greater than ten, the oil is lighter and floats on water ("light crude oil"); if

the figure is less than ten, the oil is heavier and sinks. If oil contains only about 0.24 per cent of sulphur, it is also “sweet” crude oil.

Chapter Three

The rule of oil: Why do so many of the places that have oil reserves seem to be in such a mess?

Our dependence on cheap oil has made it a "strategic asset", something to which any government that wants to develop its economy (i.e., all of them) requires unrestricted access. And this, of course, has led to some interesting political and colonial manoeuvring.

And no one has been quite as accomplished at political and colonial manoeuvring as the British – at the beginning of the twentieth century, at least. This was especially true when it concerned a commodity as valuable as oil, of which there was not one drop in

Britain. That wasn't too much of a problem, though, for a substantial and growing empire, and, as it turned out, one of Britain's colonies, Iran, had loads of oil and it seemed likely that much of the Middle East was awash with oil. Good news for Britain, but less good for budding Arab nationalists.

The Anglo-Iranian Oil Company[1] was set up to exploit the oil resources of Iran which it did under a cunning arrangement called "concessions". These were based on a very simple idea: that the country which wanted the oil (Britain or, perhaps, the USA) and which had the expertise and experience to find, extract, and refine the oil would pay a concession to the country which owned the oil because, clearly, they didn't have the expertise and so on to make use of it themselves, and nor did they have a domestic market for the oil. Britain and the USA were taking the oil off the hands of these countries, doing them a favour, almost.

So the "host" countries were paid on output: the more the foreign operating company pumped out of the ground and sold, the more money the host country received. Companies, such as Standard Oil, and governments scrambled to secure concessions across the Middle East.

Oil had become the biggest game in town. And its importance was enhanced by the changes that followed the First World War. Until the First World War, the Middle East region had been dominated by the empire of the Ottoman Turks which had existed for over 500 years. The disintegration of the Ottoman Empire in the wake of the First World War meant that the powers of the day, most notably Britain, took the opportunity to grab a bit more land, to extend their influence, if you like. Britain ended up in control of much of the Middle East and, indeed, drew new borders for countries like Iraq, Jordan (or Trans-Jordan, as it was at the time), and Palestine.[2]

This was a boom time for the oil companies. It was the 1930s and 40s that saw the agreements that would enable the exploration of the largest oil fields ever discovered, and put the Middle East slap bang in the middle of international politics.

And this process of gobbling up the Middle East was only accelerated by the Second World War. Tanks, planes, and ships needed an endless supply of oil to ensure victory. Adequate oil resources were critical, as were refining and distribution capacity. As the French industrialist and senator Henri Berenger put it, "He who owns the oil will own the world, for he will own the sea by means of heavy oil, and the air by means of the ultra-refined oils, and the land by means of the petrol and the illuminating oils. And in addition to this he will rule his fellow men in an economic sense, by reason of the fantastic wealth he will derive from oil."[3]

The war was also an important watershed in the balance of global power. The USA, now an emerging global superpower,[4] came to Britain's rescue with manpower, machines, and, most importantly, oil. The USA turned on the taps in Texas and bailed out Europe.

The huge demands placed on the oil supply in the USA meant that in order to meet their domestic and international commitments they had to begin importing oil for the first time.

And so began the USA's dependence of foreign oil. Such was the USA's concern at this that in 1945 President Roosevelt met Ibn Saud, the ruler of Saudi

Arabia (newly created in 1935) on the US cruiser *Quincy* in the Suez Canal to agree that the USA would support Saudi Arabia militarily in return for continued access to their oil through their ARAMCO (Arab-American Oil Company) concession. Indeed, as Madawi Al-Rasheed puts it in his book *A History of Saudi Arabia,* "For Ibn Saud, the meeting held great significance as he journeyed beyond his borders in search of an ally to guarantee the independence of his newly created realm. The United States was in search of oil deposits and military air bases."[5]

This "special relationship" would be essential for the USA to build its economic and military strength over the next half century.

The concessions allowed the West effectively to control supply, and therefore the price of oil. And it was the multinational oil companies which dictated terms.

OPEC

This all changed with the rise of Arab (and Persian) nationalism. While the West had carved up the region – giving countries to monarchs of their choosing – the ordinary people had become a bit fed up with being pushed around. And the producer countries had started to realize that the concessions weren't such a good deal after all, that, in fact, they were a bit of a bum deal.

So, in September 1960, Iran, Iraq, Kuwait, Saudi Arabia, and Venezuela formed the Organization of the Petroleum Exporting Countries, or OPEC, which aimed to "co-ordinate and unify petroleum policies among Member Countries, in order to secure fair and stable prices for petroleum producers; an efficient, economic and regular supply of petroleum to consuming nations; and a fair return on capital to those investing in the industry."

Clearly, the producer countries were attempting to re-assert some of the rights that they felt they had lost to multinational oil companies. OPEC members began the process of building the organization, its mandate, and its influence. By the mid-1970s Qatar, Indonesia, Libya, Nigeria, Ecuador, Gabon, and Angola had all joined OPEC.

This was bad news indeed for the multinational oil companies and their governments which had invested in their vertical integration strategy; now production would be handled increasingly by state-owned enterprises which meant, of course, that they controlled the supply and therefore the price of oil.

Oil pain

OPEC countries now produced one out of three barrels of oil, and, for a few years, everything went smoothly. The old relationships continued to work well and there were no major hiccups in supply. That is, until the 1973 Arab–Israeli war. OPEC, not noted for its pro-Israeli stance, was outraged at what they saw as Western support for Israel and flexed its new muscle. They unilaterally increased the price of oil by 70 per cent (to almost $35 a barrel) and stopped shipments to the USA and Britain. The effect was pronounced and shifted the balance of power almost overnight to the producer countries.

The price rise hurt, it really hurt: the crisis, though relatively short lived, pushed many Western economies into recession[6] and forced the USA to take some drastic measures: issuing gasoline coupons and laying plans to invade Saudi Arabia, among them.[7]

The spike did have some unforeseen side-effects though: it led to a decline in demand for oil that caused the price to collapse and the economies of the supplier countries along with it, most notably OPEC members. It also led to a whole host of initiatives from Western governments aimed at decreasing their dependence on oil.

It was at this time, for example, that that the USA introduced the Energy Policy Conservation Act, which was enacted into law by Congress in 1975, a significant part of which was the Corporate Average Fuel Economy (CAFE), the short-term goal of which was to double new car fuel economy by model year 1985. And it worked: between 1970 and 1986 "the amount of energy required to generate a dollar of wealth fell by 30 per cent."[8]

So it wasn't all bad news. That really depended on how much oil a particular country had and how developed the rest of its economy was. It also depended on the existing political climate and the degree to which ordinary people were involved in decision-making, i.e., whether the country was a democracy or not. Often, and somewhat unexpectedly, the dynamics of oil can have an unfortunate effect on democracy.

Take the example of Saudi Arabia, the world's largest producer of oil and a key ally of the West, at least for the moment. It is named after its founder and leader, Ibn Saud of Riyadh, who, in the early part of the twentieth century, consolidated his grip on the area by defeating his enemy Sherief Hussein of Mecca. He did this with the support of "warrior fanatics"[9] led by a religious zealot called Abd al-Wahhab.[10]

So, Saudi Arabia has a powerful tribal heritage based on loyalty and privilege, backed by a strong religious fundamentalism. It was not, and is not now, a fertile ground for the concept of democracy.

In addition, it has vast reserves of oil, the revenue from which goes directly into the coffers of the ruling family, or, if one were to be polite, the Saudi Treasury. This revenue reinforces the centralization of power and protects the position of the ruler as the "benefactor" of the people. Because the ruling elite of Saudi Arabia does not rely on revenue from taxes paid by working citizens, it is not really accountable to the Saudi people; instead, there is a fully fledged system of cronyism, or, as the Arabs might say, "He with the greatest *wasta* (influence) wins."

There are naturally some benefits that flow from this system. For a start there is no taxation, neither on income, nor in the form of duties, so one gets to keep one's entire salary, and the cigarettes are cheap. Wonderful! It's the reason why so many ex-pats love to work in places like Saudi Arabia.

Sometimes referred to as "rentier" states, countries like Saudi Arabia have four main characteristics:

1. Rent situations. This is based on the idea that given a resource, like oil, the state is effectively renting its income.
2. Substantial external rent. If this rent is a significant proportion of income and therefore does not require a strong domestic productive sector – for example, in Saudi Arabia, Kuwait, Libya,

Iraq, and Iran oil accounts for between 95 and 99 per cent of all exports.[11]

3. Only a small proportion of the working population is actually involved in the generation of the rent. The oil business will sustain a good few thousand local people, but it will not provide employment for a diverse workforce of millions.
4. The state's government is the principal recipient of the external rent - oil revenues flow directly into the treasury.

In essence, the wealth accumulated by rentier states does not accrue from work, but is the result of sheer chance. And, as a citizen, one receives an income simply by virtue of being a citizen. Inertia, both political and economic, can be the outcome.

Anyone who has visited the Gulf States of the Middle East will have a keen sense of the rentier economy[12] and the effect it has on every aspect of government, administration, and business.

The same is true in oil countries outside the Middle East, like Venezuela, Nigeria, and Chad, where oil revenues represent a significant proportion of GDP. The Nigerian Delta is an area of growing conflict because many local people feel disenfranchized. Likewise, in Venezuela where the old promise of "oil wealth for all" has led to mass protests by the poor and a change of government. Chad, too, while relatively new to the oil game, can expect turbulent times ahead in a country where the average wage is less than a $1 per day.

As the global appetite for oil continues to grow, governments will continue to seek secure supplies by whatever means. As Paul Roberts puts it, "Oil is not simply a source of world power, but a medium for that power as well, a substance whose huge importance enmeshes companies, communities and entire nations."[13]

China in Sudan

China's rush to secure energy resources has had a major impact on the situation in Darfur, Sudan. Indeed, many blame China and her reluctance to condemn the Sudanese government for the continuing plight of the people of Darfur; China has so far refused to agree to a UN resolution condemning Sudan.

The relationship between China and Sudan has deepened considerably over the recent past. China, as well as providing diplomatic protection, delivers billions of dollars in investment, oil revenue, and arms to the Sudanese government which is accused of genocide in Darfur and is accused by many of massacring civilians and displacing them from land they have lived on for generations to clear areas for oil production. China has a unique opportunity in Sudan as it is a country in which US companies are forbidden to do business because it has been blacklisted by Washington as a state supporter of terrorism.

Sudan is China's largest overseas oil project and China is Sudan's largest supplier of arms. Chinese-made tanks, planes, bombers, machine guns, and other arms have intensified Sudan's twenty-year-old civil war between the North and the South. While a peace treaty has been signed, conflict in the Darfur region continues unabated.

The pressure on China to find new sources of oil has increased dramatically. China is now the world's second largest consumer of oil and needs to import over 30 per cent of its oil. This figure is forecast to grow to a staggering 60 per cent by 2020. China therefore needs to secure oil supplies from wherever possible. Sudan, as an oil producer and an international pariah, represented an opportunity too good to miss for the world's fastest growing consumer of oil.

Currently, oil from Sudan makes up ten per cent of China's oil imports. Chinese workers work in the Sudanese oil fields and Chinese pipes carry the oil to Chinese ships. The China National Petroleum Corporation owns 40 per cent of Sudan's Greater Nile Petroleum Operating Company, a consortium that dominates Sudan's oil fields in partnership with the National Energy Company.

The consortium's two main fields, Heglig and Unity, produce more than 350,000 barrels per day and the China National Petroleum Corporation owns a field in Darfur, which began producing in 2005. Another Chinese company, Sinopec Corporation, is building a pipeline to Port Sudan on the Red Sea where China's Petroleum Engineering Construction Group is building a tanker terminal.

Unsurprisingly, China has been Sudan's chief diplomatic ally at the United Nations; clearly, China's oil interests have played a key role in China's diplomacy. And while the conflict in Sudan predates Chinese investment, China's pursuit of oil has dramatically increased the stakes, as well as the Sudanese government's ability to prosecute the war. Essentially, the war is a battle by the Muslim Arab government in Khartoum to control the oil resources of the south, where Christian and animist Africans live.

For many years the Sudanese government simply lacked the resources to defeat the Sudan People's Liberation Army but all that changed with increased income from oil in 1999. Sudan doubled its spending on arms and also built a number of arms factories, with the help of the Chinese. Human rights groups have also accused the Sudanese government of using oil revenues to pay for government-led ethnic cleansing of the Nuer and Dinka people who live close to the oil facilities.

As far as the prospects for peace are concerned, the leaders of the Sudan People's Liberation Army have already made it clear that should they come to power, they would force the Chinese out of Sudan as punishment for their support of the current Sudanese government. So it is not really in China's interests to pursue peace. Is China likely to risk losing ten per cent of its oil? Not really.

The USA in Africa

During the Cold War, Africa was an area of considerable interest to the USA, but in the early 1990s their interest and influence waned. That was until a number of factors – the instability of oil-producing countries in the Middle East and elsewhere, OPEC's stranglehold on world oil supplies, and the USA's dependency on oil imports – led the USA to court new and, it hoped, more compliant producer countries. This shift is most obvious in West Africa which now provides more than 15 per cent of US oil imports, almost as much oil as the US imports from Saudi Arabia. Indeed, the current Bush Administration's national energy policy, released in May 2002, stated that West Africa would become "one of the fastest growing sources of oil and gas for the American market."

The USA has been assiduously courting the oil-producers Angola, Nigeria, Chad, and Equatorial Guinea, all of which are run by regimes which are despotic to varying degrees. What is the likely outcome for the people of these oil-producing countries of the USA's new interest in the continent? Trouble, of one kind or another, seems to be the answer. In Angola, for example, oil revenues have enabled the government of José Eduardo dos Santos to build a massive army and secret police force. Apart from oil and repression, however, the country can claim little other than a booming artificial limb business.

Nigeria

Even Nigeria, though technically a democracy, has seen many of its citizens killed under the leadership of President Olusegun Obasanjo, and the unrest that has engulfed the oil region has been steadily worsening. Since independence in 1960 there has been real tension in the area surrounding the town of Warri in the Delta State, one of the most important oil towns in Nigeria after Port Harcourt. The area is home to three distinct ethnic groups: the Ijaw, of whom there are nearly ten million; the Urhobo, who have a population of around five million; and the Itsekiri, who number under a million.

The three groups claim ownership of the oil around Warri, though one group, the Itsekiri, have the most influence in the area, holding local government seats and to a large degree dictating policy. This does not go down well with the Ijaw or the Urhobo, for obvious reasons, and has created an explosive situation in the region. And while these kinds of territorial disputes have been going on for decades, oil has exacerbated them, creating a situation in which human rights abuses are rife.

Critically, the Nigerian government has not sought to intervene, either at local government or national level, to address some of the fundamental issues that trouble the region. Instead, they have resorted to force to quell unrest, and have ascribed any misdeeds to oil smugglers rather than to inter-ethnic rivalry for control of resources.

Equatorial Guinea

Equatorial Guinea, which supplies over two-thirds of its oil to the USA, is another alternative source of supply for the USA. It is a tiny country with a population of just over half a million and a history of brutal leadership and economic decline. The only thing that has saved Equatorial Guinea over the years from total collapse has been oil. It was Dallas-based Triton Energy, which had been convicted of bribing Indonesian government officials in 1993, which first found oil reserves in Equatorial Guinea in 1999 and began to form much closer ties with the government of President Obiang, who had come to power in a military coup in 1979. Since then, this small African country has seen the construction of a new US embassy and is well on its way to being removed from the list of African nations barred by the USA, on human rights grounds, from receiving trade benefits. Not that much has changed in Equatorial Guinea, mind you, apart from the oil and the wealth that accrues to Obiang and his cronies, and, of course, to Triton which receives 75 per cent of the receipts from oil - high even by developing world standard where an oil company's cut is normally closer to 50 per cent.

As the USA's economic interests in Equatorial Guinea grew, a slow shift in relations between Washington and Malabo occurred. In June 2000, the Overseas Private Investment Corporation approved $373 million in loans for the construction of a methanol plant in Equatorial Guinea, its largest progamme ever in sub-Saharan Africa. Two US companies, Noble Affiliates and Marathon, together own 86 per cent of the plant.

This political thaw has resulted from intense lobbying on the part of the oil industry, which has sought to portray Obiang as a born-again reformer, and has depicted Equatorial Guinea as a potentially huge new source of oil. It helps, of course, that the companies active in Equatorial Guinea have close ties to the Bush administration. Bush's decision to reopen the US embassy in that country was taken soon after he had received a request to do so from the oil industry.

The oil companies also worked through the Corporate Council of Africa, which represents companies with investments on the continent. In 2001, the council published a "country profile" of Equatorial Guinea, which was paid for by six oil companies and AfricaGlobal, a Washington lobby group that represented Obiang at the time. Of course the profile promotes Equatorial Guinea as a great new investment destination and hails Obiang as a great reformer, which, clearly, he is not.

In fact, Equatorial Guinea is run like a family bank, and the closer one is to "the Man", Obiang, the bigger one's share of the spoils.

Although the oil companies pay well by local standards, they have created relatively few jobs, even for a country with a population of just over 500,000. Many people living in Malabo are unemployed, and the situation is worse in the countryside where the majority of Equatorial Guinea's population lives. Many villages do not have schools, and the production of coffee, cocoa, and other cash crops has virtually collapsed. As a result, there has been an influx of people from the countryside into the cities in search of jobs, which, sadly, do not exist.

There is little in the way of political opposition to Obiang's regime as eleven of the officially sanctioned opposition parties have aligned themselves with Obiang's government in return for cash pay-offs.

Chad

The African state which has set off most recently on the road to oil riches is Chad, following the discovery of almost one billion barrels of oil in 2003. Chad's president, ex-warlord Idriss Déby, rules a country of just over nine million people with a per capita income of less than $2 per day. Oil exports are predicted to reach 250,000 barrels per day and President Déby will receive 28 per cent of the revenue that accrues (this compares with 60 per cent in Angola and 80 per cent in Nigeria). Will Chad go the way of so many oil rentier states? There are a number of interesting studies which suggest that oil, rather than being a catalyst for growth is, in fact, quite the opposite. As Lisa Margonelli points out in her book *Oil on the Brain*, "a study by Jeffery Sachs and Andres Warner showed that of ninety-seven developing countries, those without oil grew four times as much as those with oil."

In short, the benefits which should accrue from Africa's oil wealth are spreading very slowly indeed, almost as slowly as true democracy.

Venezuela

While many oil-producing countries are decidedly undemocratic, Venezuela is an exception. It is interesting, too, because of its position on the doorstep of the USA, the world's largest consumer of oil.

Oil was first discovered in Venezuela in 1921. As the Venezuelans were new to the oil game and everything it entailed, US oil companies lent a helping hand by drafting hydrocarbon laws for Venezuela, a country with the largest reserves of oil in the Western hemisphere.

The oil companies advised the Venezuelan leader at that time, Juan Gomez, an ex-farmer, the oil should be placed under the ownership of the state, to ensure that oil policy and control were neatly tied together. A national company, Petróleos de Venezuela, SA (PDVSA), was set up to extract, refine, and export the oil – with the help of US experts, of course. This model ensured that revenues from oil did exactly what had been intended: flow directly into the pockets of Juan Gomez.

Venezuela sells 70 per cent of its oil to the USA, which, in turn, accounts for 12 per cent of US oil supplies. So it goes without saying that the USA takes a very keen interest in Venezuela and anything that might interrupt their supply of oil from there. The focus, of course, is on the supply of oil, rather than on any social or democratic considerations.

The early promise that the Venezuelan government

would "sow the oil" - invest oil revenues in social programmes - has been broken many times since the 1920s. Today, half of Venezuela's population lives on less than two dollars a day, in a country which earns 42 billion dollars a year from the national oil company PDVSA.

This kind of inequity is a common feature of oil-producing countries. While oil accounts for almost half of Venezuela's GDP, only 50,000 people out of a total population of 26 million are employed in the oil industry.

Venezuela's "faith" in the wealth generated by oil has also led to an outbreak of "Dutch Disease", an economic malaise named after the collapse of the Dutch economy following the discovery of natural gas in the Netherlands in the 1960s, a situation in which resource-rich countries neglect key sectors of their economy. "Dutch Disease" in Venezuela is very apparent in food production; this once thriving sector of the Venezuelan economy now imports eggs from the USA.

Venezuela's economic woes have created social and political unrest, and a great deal of antipathy towards the USA. During the 2002 coup that forced Hugo Chavez from office it was common knowledge that the USA had supported the leaders of the coup against the democratically elected government. Chavez was returned to power after only two days, but only by successfully mobilizing the disenfranchised poor. Today, he remains locked in battle with forces supported by the USA, the biggest buyer of Venezuelan oil, to retain control over Venezuela.

Notes

1. Later to become BP (standing not for "Bloody Persians", as some have suggested, but for British Petroleum, and, more recently, Beyond Petroleum).
2. This is why the map of the Middle East has so many straight lines on it. These borders do not represent the national, tribal, or ethnic make-up of the region, but, rather, areas of Western control.
3. Tertzakian, Peter, *A Thousand Barrels a Second,* p. 40.
4. The end of the Second World War marked the end of Britain's days as an all-powerful empire.
5. p. 90.
6. Research by energy economist Philip Verleger has shown that a rise of just $15 a barrel can cause a 0.5 per cent decline in economic growth.
7. This was something that was seriously considered at the time, but was ruled out because of concern over the possible response of the then USSR.
8. Roberts, Paul, *The End of Oil,* p. 151.
9. Keay, John, *Sowing the Wind,* p. 42.
10. This was the original *Ikhwaan Muslimeen,* or Muslim Brotherhood, and the *Wahhabism* that Osama Bin Laden seems to have co-opted in more recent times.
11. *World Bank Development Report,* Oxford University Press, p. 198.
12. With one startling exception, Dubai (part of the United Arab Emirates), which has actively sought to diversify its economy.
13. Roberts, Paul, *The End of Oil,* p. 93.

Chapter Four

Carbon: What effect does burning oil have on our environment?

And that's the upside; the downside is that all the oil we're burning isn't doing our rather lovely planet any good at all.

Oil is a fossil fuel, a term for buried, combustible, geologic deposits of organic materials, formed from decayed plants and animals that have been converted to crude oil, coal, natural gas, or heavy oils by exposure to heat and pressure in the earth's crust over hundreds of millions of years. Much of the material that went into making the coal, oil, or gas that we burn today was laid down before the dinosaurs roamed, during the Carboniferous Period, about 360 million years ago.

Carbon occurs in all organic life and is the basis of organic chemistry. When united with oxygen, carbon forms carbon dioxide, the main carbon source for plant growth. When united with hydrogen, it forms various flammable compounds called hydrocarbons. Fossil fuels are hydrocarbons.

Both plants on land and phytoplankton in the sea ("plant plankton", as opposed to zooplankton, tiny creatures) are experts at converting sunlight into energy by means of photosynthesis: every green surface is full of cells that are working away making sugar as long as the sun is shining, manufacturing molecules of carbohydrates. In the very simplest terms, carbohydrates mean "energy" for humans, or, indeed, for any animal.

All energy is derived ultimately from the sun, and the ability of plants to convert sunlight into energy and to store it. Oil is really just trapped energy – trapped, buried, and heated by the earth's core. And when we burn coal, oil, or gas we release that energy in the form of heat, while releasing other chemicals into the atmosphere, most notably the gas carbon dioxide which comprises carbon, from the oil, and oxygen, from the air. Sulphur is also emitted by burning coal and is responsible for acid rain.

This is the nub of the problem. While oil and coal have been around for millennia, and have been used for centuries, they have not until relatively recently been mined and extracted and burned on such a vast scale. Does this matter, given that coal, oil, and gas are all natural substances? Emphatically, it does. It has taken hundreds of millions of years for the earth's climate to develop to the point at which it is able to support its myriad different species of flora and fauna, each perfectly adapted to its surroundings. And the earth's current climate is a delicate chemical balancing act; the achievement of just the right balance of carbon dioxide and oxygen in the air has enabled life to flourish on planet earth. How has this balance been achieved? In part, through mechanisms which absorb gases that in too concentrated a form would make life on earth impossible, among them, carbon dioxide. Carbon dioxide is absorbed by the earth's enormous forests and vast oceans; it has also been removed, or kept, from the atmosphere, by carbon being buried deep underground in the form of coal, oil, and gas.

Until about five hundred years ago the earth was doing very well, thank you, with atmospheric concentrations of carbon dioxide of only 270 parts per million. Its carbon cycle had reached a point of equilibrium where for each molecule of carbon dioxide emitted (from decaying matter and the like), one molecule was absorbed by the forests and oceans. In other words, planet earth had a neutral carbon cycle.

All that changed as humanity's appetite for energy grew. Not only did we start to burn fossil fuels - thus releasing their stored carbon dioxide - we also decimated the forests that had once helped to absorb excess carbon dioxide from the atmosphere. We had not only upset the carbon cycle; we had thrown it into reverse. As Paul Roberts points out in *The End of Oil*, "Since the late 1700s, emissions of carbon dioxide have climbed from a paltry hundred million tonnes of carbon per year to around 6.3 billion tonnes a year - about twice what the biosphere can easily absorb."[1]

With emissions increasing by 3 per cent per annum

we're on track to a figure of twelve billion tonnes of carbon dioxide by 2030, taking carbon dioxide concentrations from today's 370 parts per million to 580 parts per million by 2100.

Does this matter? Surely, even at that rate of growth, the concentrations are still minuscule? Yes, but what is becoming increasingly clear is that even current levels of carbon dioxide in the atmosphere are having a dramatic effect on the earth's temperature.

In very simple terms, carbon dioxide prevents heat from escaping from the earth's atmosphere; it produces a greenhouse effect, trapping heat reflected by the earth's surface, hence the description of carbon dioxide as a "greenhouse gas". The greater the atmospheric concentration of carbon dioxide, the more heat is trapped. And this sets up its own self-reinforcing cycle: the more carbon dioxide there is, the hotter the planet becomes; the hotter the planet, the less ice there is at the poles; the less ice there is at the poles, the less heat is reflected, and the hotter the planet becomes; and so on.

According to the UN's latest Assessment Report on climate change (produced by the UN's Intergovernmental Panel on Climate Change in February 2007), carbon dioxide concentrations of 580 parts per million, could have catastrophic effects, causing world temperatures to rise by 6.4 degrees by 2100 and leading to the extermination of life on earth. On the plus side, they reckon carbon dioxide concentrations of 450 parts per million could probably be survived.

So, the problem is serious. Even if there are contrary views – and there are a number of them – it is clear that *something* is happening. The question is how serious the consequences will be and how fast the process will accelerate. As US senator John Kerry noted, "Although President Bush just noticed that the earth is heating up, the American public, every reputable scientist and other world leaders have long recognized that global warming is real and it's serious."[2]

And the big sinners in all this? Yes, you guessed it, the "developed world", as we like to call it, accounts for over 50 per cent of all carbon dioxide emissions, and the USA accounts for half of that figure.

It is for this reason that initiatives like the 1997 Kyoto Protocol (a promise by signatory countries to reduce carbon dioxide emissions to below 1990 levels by 2012) and the UK's draft Climate Change Bill, published in March 2007, which commits the UK government to reducing carbon dioxide emissions by 60 per cent by 2050, have had such widespread support, and why they specifically target the transport sector. While the UK's industrial and domestic carbon dioxide emissions have been falling since 1974, transport-related emissions have been increasing, more than doubling since 1970.

So, while there is growing demand for oil, and powerful vested interests are encouraging this, our planet is telling us to cut back on burning fossil fuels. This paradox can be better understood by examining the future of oil reserves and looking at alternatives to our energy-hungry lifestyles.

Notes

1. pp. 124–5
2. The *Independent*, Saturday 3 February 2007, p. 2.

Chapter Five

Keep on drilling: How much oil is left, where is it, and are we going to run out?

As one would expect, there are hugely divergent opinions on the question of oil reserves: how much oil remains, and when we will reach the point when the oil begins to run out, the point at which oil production peaks. This "peak oil" point is the topic of much heated debate and controversy.

But the idea, first mooted by the Texan M. King Hubbert, has been around for over half a century. Hubbert is a legend in the oil world; his work on oil and gas reserves during the 1940s and 50s concluded that oil

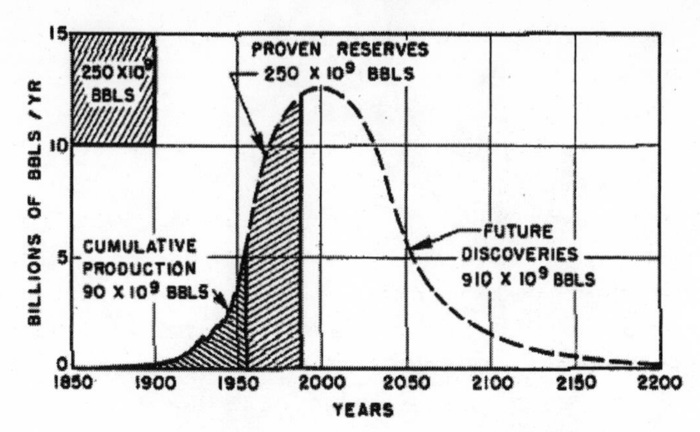

Hubbert's peak oil graph, from his 1956 paper.

fields follow a basic pattern of flow over their lifetime.

That is, when first tapped the flow of oil steadily increases to a peak of production over time as the oil is depleted. Once the peak is reached the flow reduces significantly until the field is exhausted. A simple bell curve is the result.

In 1956, Hubbert predicted that peak oil production in the USA would be between 1966 and 1972. It peaked in 1970, and, needless to say, almost everyone who had anything to do with the oil industry had thought that he was barking mad at the time of his prediction.

But this pattern of peak oil has been repeated around the world from the USA and Mexico to the North Sea. And, naturally, it has huge implications for oil supply at a time when global demand is accelerating and new finds are running at just 40 per cent of existing consumption.

So, while most accept that this is the immutable pattern of oil production, there are differing views on when the global peak will be reached, i.e., when combined global oil assets reach their peak and begin their decline.

There are two basic schools of thought on the this: the "late toppers" and the "early toppers". Essentially, the former believe that there are around 2 trillion barrels of oil left to exploit, while the latter believe the number is just over a trillion - a significant difference. And at current rates of consumption - that's without factoring in the increased demand that is predicted for China alone - we use a billion barrels of oil every twelve days or so; or, worst case, if one is an early-topper, we have about thirty-three years of oil left at current rates of use.

But that's fantasy, isn't it? No serious commentator believes that demand for oil will abate; rather, as discussed earlier, demand is likely to rise at something close to 2.5 per cent per annum, and at this rate we will be burning through a billion barrels of oil every six days by 2032. So, at best, the world's supply of oil is likely to last for twenty years or so.

The early-toppers believe that we've pretty much found all the oil we're going to find and that there simply isn't any significant reserve of oil yet to be discovered. In other words, even at our current rates of consumption, a "good" find with reserves of say a billion barrels is only, as noted previously, twelve days' worth of oil - an alarming prospect.

The early-toppers maintain that there simply aren't any super-fields like Ghawar in Saudi Arabia or Canterell in Mexico[1] left to find, that the oil companies, despite having been exploring for decades, are still only replenishing around 40 per cent of the oil used each day with new discoveries. As Paul Roberts puts it, "since 1995, the world has used at least 24 billion barrels of oil a year but has found, on average, just 9.6 billion barrels of new oil annually."[2]

From the late-toppers perspective things don't look quite so bad: a remaining 2 trillion barrels, even allowing for growth in demand of 2.5 per cent each year, would still give us around fifty years of oil.

For obvious reasons, it's important to try to understand who is right: the early- or late-toppers. So, let's start with our existing oil reserves, fields that are pumping the majority of our oil at the moment. That means looking at the claimed reserves of the OPEC cartel. Until the mid-1980s, OPEC reserves stood at around 400 billion barrels. They then decided that each country's quota – the amount of oil it could pump and sell – should be determined by its proven reserves. A good idea in principle, perhaps, but, in practice, every single OPEC member simply "re-evaluated" their reserves between 1985 and 1986 increasing their "proven" reserves from a total of a little over 400 billion barrels to more than 700 billion barrels.

Arguments, such as "We've found new ways of measuring" and "We can get more out than we thought" were advanced, but it all seemed a bit fishy. And, naturally, the oil companies weren't going to question these new estimates, as that simply wouldn't be in their interests.

To be fair, some of the arguments raised did have some validity. "Enhanced recovery techniques" centre on increasing pressure within the oil field in order to force more oil out.[3] These techniques include the use of pressurized water or steam. So OPEC could claim some increments due to these techniques, but, at best, those would amount to an additional 20 per cent. And, by my calculations, that amounts to extra reserves of only a little less than 300 billion barrels.

It could also be argued that once enhanced recovery is required (i.e., at the point at which the "natural" flow slows and, therefore, daily yields fall) the field has already reached its peak. As Donald Coxe pointed out in a 2005 report for the Bank of Montreal, "Isn't water-flooding petroleum Viagra for aging wells?"[4]

If so, we're in trouble. But, for the time being, 700 billion barrels of OPEC oil is the number that every government and oil company is working with.

The other argument that the late-toppers put forward is that while there may be a slowdown in the discovery of conventional oil fields there is plenty of "unconventional" oil around. This includes the tar-sands of Alberta in Canada which alone hold a trillion barrels of oil, it is claimed. While that is an impressive quantity of potential oil, companies mining it would have to be able to accomplish two things: firstly, they would have to be able to produce more than a million barrels per day; and, secondly, they would have to produce a barrel of oil for something like the cost of a conventional barrel of oil. They would also have to take into account the environmental impact of the production process.

It seems that the only people talking up the late-topper arguments are either oil companies, governments with oil wells, or special interest groups like car manufacturers. The vast majority of independent analysts paint a more conservative, early-topper picture.

So, where is all the oil? Well, most of it, almost 65 per cent, is in the Middle East and well over 70 per cent is controlled by OPEC.

World oil reserves

Country	**Reserves** (in billions of barrels)
Saudi Arabia	265
Iran	133
Iraq	115
Kuwait	102
United Arab Emirates	98
Venezuela	80
Russia	60
Libya	39
Nigeria	36
USA	21
China	18
Qatar	15
Mexico	3
Algeria	11
Brazil	11
Kazakhstan	9
Norway	8
Azerbaijan	7
India	6
Top 20 countries	1,047
Rest of world	158
Total	1,205

Source: *Oil & Gas Journal*, Vol. 103, No. 47, 19 December 2005.

There is no getting away from the fact that the Middle East and OPEC will continue to dominate world supplies of oil: no matter how hard the USA and other countries try to develop non-OPEC supplies, they will remain dependent on OPEC for oil.

And that would not necessarily be a problem if the oil heartland were a stable region, but it is not. Saudi Arabia, which holds over 25 per cent of the world's oil, is profoundly unstable. Its economy is totally dependent on oil and it is torn between being a monarchy and a theocracy. The resurgence of that very particular brand of Islam, Wahabbism, is likely to make for some interesting times both domestically and internationally. It is no coincidence that fourteen of the eighteen hijackers involved in the attacks in the USA on 11 September 2001 were Saudi citizens. And two Gulf Wars, one essentially ongoing, have been fought for oil - a slightly simplistic take, perhaps, particularly given the neo-conservative ideology of the present White House, but it would certainly be fair to maintain that neither Kuwait nor Iraq would have warranted such direct action, had oil not been involved.

So, given both diminishing reserves and the huge price tag of oil - both economic and moral - that comes with our dependency, isn't there another way to power the world? Keeping on drilling can't be the only way in the long run as there simply aren't sufficient reserves; and, even if there were infinite quantities of oil, we would still be killing the planet if we were to continue burning oil as we do currently.

Notes

1. Ghawar is the world's largest and Canterell the second largest oil field.
2. *The End of Oil*, p. 51.
3. Conventional recovery techniques are able to extract only 30 per cent of the oil in any given field.
4. Coxe, Donald G. M., Bank of Montreal report, "Big Footprints on the Sands of Time, and Little Footprints of Fear", 30 March 2005.

Chapter Six

The DTs: Will we cope without oil?

The good news is that there is another way; we *can* cope without oil. In fact, there are loads of alternatives to using oil at the current rate; from bio-fuels and renewable energy to simple conservation.

Most of the oil used today is for transport, and half of that is burned by cars. So, a look at what's happening in the car industry should give us a better idea of potential alternatives. Although there has been much discussion about alternative power for cars, there are really only two alternative fuels: hydrogen and ethanol.

Ethanol has been around for longer and is already in

use. Like methanol, it is produced from crops, such as sugar cane or maize, or anything that can be grown and harvested on a significant scale. This is why ethanol is labeled a "bio-fuel". Once harvested the crop is cooked, fermented, and distilled to create a flammable spirit, not unlike cheap vodka.

One really useful quality of ethanol is that it can used in conventional cars (with a few simple modifications to the car's engine). It would be possible to use existing petrol distribution and retailing infrastructure to get ethanol to market, and it is far less polluting than petrol. So why don't we just switch over and save everybody all this hassle with oil? A good question. The answer is that to grow sufficient crops to provide the ethanol required by current demand (an estimated 42 million barrels a day) would require a continent or two dedicated to growing them, and that's just to fuel cars.

In Brazil, which adopted ethanol with great zeal in the mid-1970s in response to the oil shock of 1973, 40 per cent of petrol usage has been replaced with ethanol. But Brazil is still the fourth largest producer of carbon dioxide in the world, so switching to ethanol is clearly a far from perfect solution. The key problem is the sheer volume of crops needed to produce sufficient quantities of ethanol and the impact that such large-scale farming has on the natural environment. Indeed, many commentators believe that Brazil's relatively high emissions of carbon dioxide are the result of rain forest destruction and the loss of capacity to absorb carbon dioxide. That deforestation can be blamed in part on the demand for land on which to grow crops for ethanol.

Evidence like this, however, has not stopped the USA from seeking to increase its own use of ethanol. The

White House has even committed itself to substituting 20 per cent of the petroleum it uses for ethanol by 2017,[1] so it's boom time for ethanol.

Hydrogen has also come a long way over the past few years. Indeed, the early 1990s were a time of huge interest and investment in the hydrogen cell, a sophisticated battery that delivers large amounts of electrical power. Car makers like Daimler were promising to spend over a billion dollars on hydrogen fuel cell development in order that it could "compete against the internal-combustion engine."[2]

Sadly, this stirring promise has received a bit of a reality check. Creating a whole new power plant, an electric engine powered by a hydrogen fuel cell, and using this to power a car that would be as cheap to buy and run as a conventional car, was always a big ask. The day when manufacturers will reach this mass production goal is still some way off; any economies of scale are extremely unlikely in the short- to medium-term.

However, for many people, including most of the car industry, hydrogen or liquefied petroleum gas (LPG) are the future. LPG is the only fuel that can act as a "bridge" between our existing oil habits and a leaner, less oil-intensive future. To begin with, it contains more hydrogen and less carbon, so it is both better for the planet and also more easily refined into hydrogen (for fuel cells). LPG would also allow us to continue to use some of our existing refining and distribution assets.

More successful to date have been hybrid cars which use both electric power and a petrol engine. While the car is powered by the petrol engine, it charges batteries that provide the car with power when not too much is required, when driving in town, for example. Hybrid cars have been successful commercially, but still represent a very small part of the market, largely because they tend to be more expensive than conventional cars.

To make things even more challenging, existing car manufacturers spend billions of dollars a year on the internal combustion engine. Decades of development mean that they rarely break down, and they fit easily into just about any vehicle that car manufacturers choose to design. The internal combustion engine is the most reliable power plant that has ever been produced: tough competition for any new market entrant.

So there are some encouraging signs in the car market, but it seems clear that unless the true cost of car use is built into the price of purchase and use, consumers will not readily switch to a less economically damaging, less oil-dependent alternative. Most consumers, when about to pay £30,000 for a car are more interested in whether the seats are leather than how much a tank of petrol will cost or what the ecological consequences of burning that tank of petrol will be. This has led some to argue that the environmental cost of car ownership should be built into the price. By some estimates, even the most efficient petrol engine accounts for over two thousand dollars' worth of environmental damage over its lifetime. If you include the cost of "securing" oil in places like the Middle East, that figure increases to over three and a half thousand dollars[3] – a significant surcharge, indeed.

As far as renewable sources of energy like the sun,

wind, and water are concerned, for obvious reasons, these do not seem to offer much help with future transport, although solar power has been used with some success to power small cockroach-shaped cars.

That leaves the last option of conservation. Driving less, in more efficient cars, may catch on. Indeed, it is probably the one thing in which we can participate directly and immediately that would make a significant difference to our oil consumption and dependency. Imagine if the USA were able to increase the efficiency of its cars by just 25 per cent; oil imports could be reduced by almost three million barrels a day, a hugely significant figure. However, such a change would require a huge shift, not just in consumer attitudes but in government policy, too, which, over the last decade has achieved more or less exactly the opposite and brought about a boom in oil consumption.

To quote Paul Roberts, "Nearly every participant in the modern energy economy, from individual consumers to multinational oil companies to superpowers, is so deeply invested in the status quo that any fundamental change poses enormous political and economic risks."[4]

Fuel cells

Fuel cells have actually been around for a long time – since 1839, in fact. Sir William Grove knew that water could be split into hydrogen and oxygen by sending an electric current through it (a process called electrolysis). He thought, quite rightly, that if you could reverse the process you could produce electricity (and water), and he went on to prove as much.

In the past decade, however, there has been renewed interest in fuel cell technology. In 2003 President Bush announced a programme called the Hydrogen Fuel Initiative (HFI) during his State of the Union address. This initiative, supported by legislation in the Energy Policy Act of 2005 (EPACT) and the Advanced Energy Initiative of 2006 aims to develop hydrogen, fuel cell, and infrastructure technologies to make fuel cell vehicles practical and cost-effective by 2020. The USA has spent over one billion dollars on research and development so far.

So what is a fuel cell? What does it do? And how does it work? In essence, fuel cells produce electricity without any of the nasty side effects which result from burning fossil fuels, the only by-products being water and heat. Unlike a conventional battery, a fuel cell will not go "flat". A conventional battery has a fixed store of chemicals and once it has converted those chemicals into electrical power, the battery has either to be recharged or discarded.

With a fuel cell, on the other hand, chemicals flow into the cell so it doesn't go flat; as long as there are chemicals available to it, the cell will produce electricity.

The polymer exchange membrane fuel cell (PEMFC) is probably the front-runner for powering cars, trucks, and perhaps homes in the future. It has four main components: the anode (the negative part containing the hydrogen), the cathode (the positive part containing the oxygen), the electrolyte (the membrane that only conducts positive particles - i.e., the oxygen - and therefore blocks negative ones - i.e., the hydrogen), and a catalyst (which facilitates the reaction between the oxygen and the hydrogen and is usually made of platinum).

Electricity is generated as the hydrogen is forced through the catalyst and split into particles (ions and electrons) that, as they move through the cell, are harnessed to generate electricity. This reaction in a single fuel cell produces only about 0.8 volts which is not a great deal, so, in order to increase the output, the cells are "stacked".

Once the stack is generating sufficient electrical output, or power, this has to be transformed into the kind of mechanical power that would drive a car, for example. This is done by means of an electric motor which connects to the conventional components of a car such as the transmission and so on.

So far, so good. And the picture looks even rosier when you consider the relative efficiency of the fuel cell as opposed to the internal combustion engine. A fuel cell converts almost 80 per cent of the energy from the hydrogen it uses into electrical energy, and 80 per cent of that output is succesfully harnessed to do mechanical work. That is an overall efficiency of more than 60 per cent. The internal combustion engine is a very shabby performer by comparison, giving an efficiency figure of only a little over 20 per cent.

However, there are still some significant problems with fuel cells.

Firstly, their cost. They are very expensive to make, one reason for which is that the best catalyst is platinum, a very expensive metal.

Secondly, their durability. Because the cells contain a lot of water, they don't cope very well with extremes of weather. Their vulnerability to freezing, when the weather is cold, and to evaporation, when it is hot, are still major problems for the fuel cell.

Thirdly, their range. Three hundred miles would seem a reasonable distance to cover on a tank of petrol. To achieve that range with a hydrogen fuel cell would require a driver to tow a mini-fuel tanker behind his or her car, or a spacious interior capable of seating five comfortably would have to be replaced with something resembling a very snug, single-seater fighter cockpit.

Finally, fuel cells require a lot of energy to produce. While hydrogen is the most common element in the universe, it does not, unfortunately, exist in its elemental form on earth. So to create elemental hydrogen we have to extract it from compounds that contain it, such as water or fossil fuels. Just creating the hydrogen necesssary to produce fuel cells requires a considerable amount of energy.

Naturally, this makes the fuel cell car a little pricy and matters are made worse by a current lack of infrastructure for the generation and delivery of hydrogen. Establishing such an infrastructure in the USA alone is likely to cost more than half a trillion dollars.

However, fuel cell technology does hold out the possibility of a true alternative to fossil fuels, both in terms of pollution (as the only by-product of fuel cell operation is pure water) and as a way of breaking our dependence on oil.

Resource wars

In 2006, John Reid, the British Defence Secretary, warned that our diminishing natural resources and climate change could lead to conflict to protect resources such as oil, water, and viable agricultural land. Indeed, Reid cited Darfur as an example of just such a conflict; "a warning sign", he called it.

With the world's population rising, global consumption soaring, energy supplies declining and climate change accelerating, we are literally on the verge of a global struggle for resources. And the separate camps that are now emerging, both religious and national, may well prove to be the beginnings of future global power blocs.

Indeed, if some of the more extreme climate change predictions are proved correct, we may find ourselves thrown into armed conflict. Such predictions warn not of a gradual warming of the planet that leads to a slow rise in temperatures, but of a climatic tipping point when the natural regulatory mechanisms of the planet break down irretrievably and widespread devastation results, on a scale with which humanity is ill-equipped to cope.

It is important to understand that there are likely to be significant social consequences of climate change. Climate change will not only bring about the degradation of ecosystems; it is also likely to result in the disintegration of entire human societies.

One of the best examples of how countries are preparing for this scenario appears in Michael T. Klare's book *The Coming Resource Wars,* in which he describes the USA's involvement in the Caspian Sea area and the joint exercises undertaken with Kazakhstan, Uzbekistan, and Kyrgyzstan over a decade ago. These were significant because they marked the beginning of a US presence in the region in the wake of the collapse of the Soviet Empire. But why was the USA engaging in exercises in that particular area? Oil, is the answer as there is believed to be over 200 billion barrels of it beneath the Caspian Sea.

The USA was hedging against disruptions to its supply of oil from elsewhere, trying to ensure that it would have unfettered access to oil and gas in the region. To achieve this, the USA is willing to spend billions of dollars on military support, or, as President Clinton pointed out at the time, "By working closely with Azerbaijan to tap the Caspian's resources we not only help Azerbaijan to prosper, we also help diversify our energy supply and strengthen our national security."[5]

National security was something that Jimmy Carter was also very concerned about during the Iran–Iraq war when Iran was threatening the flow of oil through the Persian Gulf. At the time, the US government did everything in its power to ensure the oil kept flowing including re-flagging oil tankers (making them protectorates of the USA) and patrolling the Gulf with warships. The USA has always been prepared, and remains so, to use whatever means necessary to protect its vital interests.

While US foreign policy was focused on the Soviets and their containment until 1990, it is now focused on vital resources and their control and protection. And oil is the most vital resource for the most developed economies. In other words, foreign policy has become a matter of economics, not ideology.

This "econocentric" approach became official policy from 1993 under President Clinton who made clear that the USA's economic and security interests were inextricably linked.

The difference between the world views of George W. Bush and Al Gore lies largely in the different ways in which they seek to apply the econocentric model when it comes to energy policy. While it seems clear that George W. Bush favours securing – at virtually any cost – traditional resources, such as oil, from overseas to keep the US economy on track, Al Gore favours pursuing alternative energy options: renewable sources of power, and so on.

Certainly, in the short term, the world seems to be locked into a contest for vital resources, which has the potential to become a real conflict, particularly as the demand for increasingly limited resources builds as a result of the growth of the economies of countries such as China and India. The stakes are high, and even without the environmental consequences of global warming, there is a real risk that the drive to secure dwindling global resources will lead to serious conflict.

Power for the people

Back in the old days we were all pretty much self-sufficient. As hunter-gathers we collected our own fire wood and cooked up rabbit stew (or whatever we were eating) without the aid of a national power infrastructure. Indeed, we managed to get by with only our own power supplies right up until the advent of electricity, so not that long ago, really.

Once we'd invented the electric light bulb, we quickly developed all sorts of uses for electricity in the home and at work. Just about everything runs on electricity. Even gas- or oil-fuelled central heating requires electricity to run the pump (and these also burn fossil fuels).

As our populations grew along with our urban centres, we built more and more power stations to generate electricity. This was supplied to our homes via a national power grid and the amount of electricity we used was measured using a meter. That's just the way it's been for about as long as most of us can remember, and, in some respects, it's a good system, certainly better than the coal fires that everyone used for heating and cooking in the early twentieth century, the smoke from which helped to create London's infamous "pea-soupers", foul mixtures of smoke and fog.

We still burn fossil fuels - mainly coal or gas - to generate the electricity that we now use for cooking and heating, but on a massive scale, in power stations, making steam to power turbines which produce electricity.

The utility companies responsible for electricity generation forecast demand for electricity and generate it accordingly, responding to peaks by burning more fossil fuel in more power stations, or "utilizing spare capacity".

This centralized, national production model is extremely resource intensive. First, a multi-million-pound power station must be built (ideally, not too close to anyone's home); then, it must be supplied with millions of tonnes of coal - most probably by train - and then the electricity generated is transmitted via enormous cables suspended from pylons to substations, from where it is transmitted to homes and workplaces. This is what is known as the National Grid. As contented consumers of electricity, just about the only time we are aware of all this is when we receive our electricity bill, or when there is a power cut.

Generating power in this way, however, produces about 75 per cent of our global carbon dioxide emissions. So around three quarters of our emissions of the most significant greenhouse gas result from the generation of electricity in power stations, and global demand for electricity is growing. China plans to build sixty new coal-fired power stations every year for the next decade just to keep up with its demand for energy.

The economic model for a traditional coal-fired power station allows up to about thirty years for the capital investment to be repaid, so there is a lot of money tied up in existing assets that investors will not see a return on for many years. This situation is exacerbated by the fact that older power stations, which have already repaid their capital investment, provide the cheapest electricity because there is no longer any capital cost associated with running the power station, only the cost of maintenance and the coal. In the USA, for example, some older power stations can produce electricity for as little as two cents a kilowatt, in other words, extraordinarily cheaply.

But the older coal-fired power stations are the worst polluters as they do not have the benefit of more recent "clean-coal" technology (filtration and cleaning techniques that remove chemicals like sulphur and carbon dioxide from coal smoke) that is mandatory for any coal-fired power station built now. This, of course, presents something of a dilemma in a highly competitive energy market in which the cheapest producer gets the business. There is a lot of pressure for the older power stations to continue to produce low-cost energy, particularly because energy, after all, means economic growth.

This dilemma is one of the reasons why the US administration decided to pull out of the Kyoto Protocol. It felt that the USA would be unable to meet its commitment to reduce carbon dioxide emissions because of its existing power-generation infrastructure. Even the possibility held out by Kyoto of carbon trading (whereby countries which emit large quantities of carbon dioxide can "offset" their emissions by buying a "right to pollute" from countries with lower levels of emissions) would not have helped sufficiently, and, of course, the coal industry lobby in the USA has a long track record of successfully influencing government policy.

What are the alternatives to fossil fuel-powered power stations? Power generated by nuclear fission has been a popular choice in some parts of the world, but nuclear power stations not only make coal-fired power stations look cheap to construct and maintain, but they also have a generally poor public image. They produce significant quantities of highly radioactive nuclear waste that remains dangerous for years to come and nobody has yet come up with an absolutely compelling solution to this potential hazard. Even though the waste issue has largely been overcome with new fast-breeder reactors, which re-use spent nuclear fuel, concerns over the overall safety of nuclear power persist.

There is another nuclear power option which is to utilize nuclear fusion rather than fission (atoms produce energy in the process of fusing together rather than in the process of splitting apart). This is the reaction that creates the sun's energy so it is certainly effective.

The other alternatives to fossil fuel power centre on what are known as "renewables": sources of energy that don't require anything to be burned and thereby destroyed. The king of renewables, at least in terms of power generation, is hydroelectric power, generated by the force of running water, which accounts for around 7 per cent of global power production. But, as a rule, hydroelectric power requires a large dam and a lot of land to be flooded in order to generate a significant amount of power. So while hydroelectric power has proved successful in many places in the world, building more large dams is not necessarily the best option for cleaner energy, and is, in any case, impossible in drier areas.

The other renewable sources of energy, solar and wind power, are mere minnows at this stage, accounting for just half a per cent of global power generation. And while both solar and wind power are becoming increasingly effectively utilized as power sources, they are both intermittent sources of power. Even on the windiest hill, there will be times when the wind does not blow, and on even the sunniest day, the sun will set. So if one wants to have a reasonable supply of power from either the sun or the wind, one has to build more capacity than is required. For example, if one wants to produce 100 megawatts of power by harnessing the power of the sun or the wind, one has to install sufficient solar panels or enough wind turbines to generate 500 megawatts of electrical power.

On the other hand, both wind and solar power generation are quick to plan, build, and install. A wind turbine or a field of solar panels can be up and running within a year, and, while there are some manufacturing and installation costs, once solar or wind-power installations are functioning they require minimal maintenance and absolutely no fuel. From an investor's point of view, too, that has to be good news.

Such is the potential for wind-generated power that some have predicted that it could be supplying almost 15 per cent of the global demand for power by 2020. That prediction is based, however, on the centralized model of power delivery and many argue that it is the fact that power supply is so centralized in the first place that is a very significant part of the problem. They make the simple point that it makes more sense, both environmentally and economically, for each of us to generate our own electricity.

This is a strong argument. If one considers that most contracts for the generation and supply of power are awarded on a cost basis, i.e., the cheapest source of power wins. The efficiency with which the power is generated and distributed is not taken into account. An investor wants to know when he or she will get his money back, not, for example, what percentage of the power generated is lost through poor conductors. The issue is similar to the way in which people generally buy cars, choosing by model and specifications, not by the cost per year to fill a car with fuel.

When was the last time any of us needed 280,000 kilowatts of electricity? Never. But that is how much power a nuclear power station produces. We simply don't need that kind of power to make toast in the morning, or to boil a kettle.

Renewables have helped to start the self-generation revolution. They are quick and relatively easy to install on a house and will only become more accessible, more efficient, and more affordable as the demand increases and economies of scale begin to take effect.

Notes

1. "The Big Green Fuel Lie", The *Independent,* 5 March 2007.
2. Koppel, *Powering the Future,* p. 222.
3. Roberts, Paul, *The End of Oil,* p. 274.
4. Roberts, Paul, *The End of Oil,* p. 274.
5. Klare, Michael T., *The Coming Resource Wars,* p. 16.

Chapter Seven

Rehab: How bad can it get?

How bad things will get depends on a number of factors, particularly important present-day trends – political, social, economic, religious, and environmental – which have a bearing on our energy policies and use today. We need to explore the possible impact of those trends in the future. Let's begin with religion.

Jesus versus Mohammed

We don't like it when people try to kill us. It doesn't really matter where we come from, what we believe in or where we are in the social pecking order, we don't like to be shot at, bombed, or attacked in any other way.

As a general rule, that kind of behaviour tends to provoke a response of some kind – retribution, in a word. As the Old Testament of the Bible puts it: "a life for a life, an eye for and eye."[1]

But the world is already heavily involved in a cycle of attack and counter-attack in the part of the world which has the most oil, the Middle East. What is officially known as the "War against Terror" is being fought in Iraq, Israel, Palestine, Lebanon, and Afghanistan, with the prospect of being extended into Iran and Syria.

The USA and its allies are in these countries, we are told, because they are home to bad people who want to kill us, people like the murderers who flew airliners into the World Trade Center and the Pentagon in 2001. These bad people go by the name of Al-Qa'ida, and their very bad leader is Osama Bin Laden. These various wars have nothing to do with oil, supposedly, and everything to do with the security of the "Free World". In their eagerness to prosecute this "War on Terror", the USA, the UK, and their allies have killed hundreds of thousands of people, the overwhelming majority of them Iraqis,[2] Afghanis, Lebanese and Palestinians.

The USA and its allies have helped to create a strongly polarized world, divided by radically different beliefs. Worse still, they have forced everyone to choose sides, to be come radicalized, to join one or other "club". On the one hand, Muslims are driven to support Al-Q'aida; on the other, as George W. Bush likes to say of his moral crusade against terror, "If you are not with us, you are against us."

This radicalization of an inherently unstable part of the world, has a crucial influence on our future. The growing antipathy of the people of the Middle East to the West can only further impede the West's access to oil in the region. Indeed, one of Bin Laden's stated aims is to take control of the Saudi oil fields.

The Muslim world understands very well that what they term the "Christian West" owes its military and economic power to oil. It is not only the West, however, that will be contesting access to declining reserves of oil. China, as its economy continues to grow and its own, limited reserves of oil dwindle, will be increasingly driven to seek oil through alliances with key oil-producing countries in the Middle East.

Russia, although it has enormous oil and gas reserves, has already made it clear that, regardless of commitments to supply customers outside of Russia, its own needs will come first; and Russia has a nuclear arsenal to keep potential poachers at bay.

Increasing demand for oil in the face of declining reserves has the potential to destabilize large areas of Africa and South America, too. Already, oil-producing states in these regions manage a precarious balance between international demand for oil and domestic development; most of them would be easy targets for oil-hungry predators. Attempts to intervene on either continent have the potential to create another Middle Eastern-style scenario and the consequences could be devastating because of Africa's inherent political and social volatility and South America's proximity to the USA.

There is almost limitless scope for things to go very badly wrong indeed, and, even worse, there's no good reason to suppose that they won't. However, the challenge to devise a new framework of international relations which deliver peace and stability at the same time as securing access to energy supplies is not insurmountable. Admittedly, those new international relations would be very different from those that have held sway over the past few decades, which seem to have been more focused on securing energy supplies at any cost - by bribing, cajoling, and/or threatening any non-compliant government.

Cooking nicely

The doomsday scenario of worsening wars would most likely bring with it an acceleration of climate change; after all, countries with wars to fight are unlikely to prioritize reducing their carbon footprint. Despite the best efforts of scientists using increasingly complex predictive models, the specific, regional impacts of climate change are notoriously difficult to predict accurately. Nobody really knows exactly what is going to happen in particular parts of the world, or when. What is clear, though, is that the most extreme forecasts don't bear thinking about. The future, should the governments of the world take only ineffectual action to mitigate global warming, could well be catastrophic: widespread droughts, worse famines, flooding, more powerful hurricanes, and so on. Such environmental catastrophes would undoubtedly, as we have already seen to a limited extent in the USA following Hurricane Katrina in New Orleans, result in civil unrest and economic chaos.

These kinds of consequences are, in fact, already apparent in some areas of the world, most notably in Africa. For decades, the ownership of the Bakassi Peninsula between Nigeria and Cameroon has been disputed, and it has been at the centre of several conflicts in 1981 and the early 1990s. These conflicts have revolved around the environment: rich fisheries and, particularly, oil reserves in the waters surrounding the peninsula have heightened tension between the two countries.

Lake Chad, in the Bakassi region, has been severely affected by drought and desertification. It has decreased in size from an average of 4,000 square miles in the dry season in the 1960s to an area of only 839 square miles today. As a result, fishermen have been displaced and forced to find somewhere else to live; this has put enormous social and economic pressure on the region and led to localized conflict. While it remains unclear whether this decrease in the size of Lake Chad is being exacerbated by global warming, it is clear that there is a link between the increasing demands of an expanding population and the dramatic shrinkage of one of Africa's largest freshwater lakes over the past forty years.

Basically, overgrazing around the lake reduces vegetation, which compromises the ecosystem's ability to recycle moisture back into the atmosphere. This results in fewer monsoons, which used to replenish the lake, and, consequently, droughts, which, in turn, mean that more water is taken from the lake to irrigate crops. At the same time, the Sahara Desert has been edging southwards, exacerbating the desertification of the region. Against this backdrop of acute environmental pressure, competition over oil resources is a catalyst to worsening conflict.

A similar, though infinitely worse, scenario is playing out in the Darfur region in the west of Sudan. The root of that conflict lies in attempts by the ruling Arab elite in Khartoum to drive non-Arabs from the land in order to control resources. Displaced people place huge pressure on the already fragile balance of social and economic wellbeing in surrounding areas, resulting in localized conflict. China's interest in the region's oil resources and its investment there have made matters worse by blocking international attempts to tackle the problem.

Lake Chad and Darfur may seem far-away places and conflicts to many readers, but they serve as warning signs for the future. Environmental degradation, whether exacerbated by the burning of fossil fuels and global warming or not, has a critical impact on communities and their ability to survive.

That's expensive

It might also help to convince the oil market that $50 a barrel is really very little to pay for oil. Frenzied traders would begin frantically to buy and sell oil, and prices would soar.

Indeed, Stephen Leeb in his book, *The Coming Economic Collapse: How You Can Thrive When Oil Costs $200 a Barrel*, believes there could be a quadrupling of the oil price. At that rate Leeb foresees an energy crisis that could spell the end of modern civilization, touching off hyperinflation, bringing about double-digit interest rates, and causing the world's economy to collapse.

Investment strategies would change dramatically. People would move out of inflation-sensitive bonds, stocks (especially small-cap stocks), and move into inflation-proof treasury bills, gold and gold stocks, oil and oil shares.

There is a lot of evidence related to previous "oil shocks" to support this argument, most notably the 1970s oil shock when prices rose by up to ten times, resulting in hyperinflation and double-digit inflation.

Food production would fall dramatically as oil-intensive farming methods became untenable. One possible result of this might be the introduction of a Second World War-style rationing system. This provided each person with a frugal diet (twelve ounces of bread, six ounces of vegetables, a pound of potatoes, two ounces of oatmeal, an ounce of fat, and six-tenths of a pint of milk per day, supplemented by small amounts of cheese, pulses, meat, fish, sugar, eggs or dried fruit).

Rationing would not be restricted to food, however. Cars would spend a lot more time parked than being driven as the idea of driving to work would become daunting, with a seventy-mile round trip in a car that did 35 miles to the gallon costing almost £60 a day, or £300 a week, just for the petrol.

In turn, the suburbs would lose their allure and house prices would fall dramatically. Our current energy-dependent lifestyles would be gone forever.

Parachute

If the coming meltdown proves to be unavoidable, the best we can hope for is that we will be able to manage the process. This would mean that society would have to start a deliberate and well-planned process of simplifying its structures and reducing its reliance on non-renewable energy. In other words, we'd have to turn an economic model based on continual growth into one to accomodate contracting economies. That's a big ask, but, as Donella Meadows points out in her book, *Beyond the Limits: Confronting Global Collapse, Envisioning a Sustainable Future*, it is as difficult for someone now to envisage the world that might evolve from a sustainability revolution as it would have been for an English coal miner of 1750 to imagine a Toyota assembly line. Like other great revolutions, a sustainability revolution would result in great gains, and great lossses, and change the face of the land, our institutions, and our cultures. Like the other revolutions, it would also take centuries to develop fully. It is already underway, however, and the next steps need to be taken urgently.

Similarly, Howard and Elizabeth Odum in their book, *A Prosperous Way Down: Principals and Policies*, take the view that there are precedents in ecological systems that suggest that global society can "descend" prosperously, reducing assets and population in line with decreasing available natural resources. What is required, however, for society to pull together is that everyone should understand that society as a whole needs to adapt to making do with less.

In other words, we are living in the twilight of a great Age of Luxury, and everything about our lives is about to change: our homes, which sources of energy we use and how we use energy, our diet, and the ways in which we get around. All this change could be overwhelming. It is perfectly conceivable that we will end up fighting for our survival against implacable rivals for dwindling resources while trying to cope with an energy crisis and environmental meltdown. Our common future depends very much on the choices we make now, and the commitment with which we act on the decisions we make.

Notes

1. Exodus 21: 23–27.
2. Brown, David, "Study Claims Iraq's 'Excess' Death Toll Has Reached 655,000", *Washington Post*, 11 October 2006, p. A12.

Summary

The big question for all of us is not how much *oil* we need, but how much *energy*, because it is our growing demand for energy that has driven our dependence on oil.

Oil has created enviable wealth, enabling us to enjoy luxurious lifestyles, but that luxury has cost us dear, not only environmentally, but socially and politically, too. It seems that it is fine for Saudi Arabia to continue to torture people, just as long as it continues to supply oil, and does anybody really care about the plight of Chad, or Venezuela, as long as it is possible to buy petrol for less than a pound a litre?

As participants in all this we can choose to have a voice, or we can simply pretend that we don't know what is going on. Most of us choose the latter, because we're so tied in, so enmeshed, that there doesn't seem to be anything we can do. Well, that's changing. Climate change, which is already happening, and will continue to happen for many years regardless of whatever we do right now, is forcing us to think about what we're doing with oil, and why we're doing it.

Oil has become so critical to our transport systems that any threat to it has prime ministers and presidents reaching for the red phone. We simply can't function without it.

To add to the importance of oil, it is clear that it is running out; that is a question of when, not if. It is also clear that competition for our declining oil resources is intensifying. The days of the West dictating terms in the Middle East and elsewhere are long gone. China is poised to become the new power-broker in the Middle East, and perhaps the world as a whole.

On a personal level, we're scared that our lives will change, for the worse rather than the better; afraid, for example, that we won't be able to fly to Spain for a summer holiday. Governments are scared of how their people will react if they can't keep the price of petrol at the pumps in line.

How will all of this play out over the next few years and decades? That's an important question, and one to which there is no easy answer. There are two basic scenarios: the first is that everything will go horribly wrong, resulting in a downward spiral of war, disease, and death; the second is that progressive solutions will allow for a more positive prognosis.

The "horribly wrong" scenario

This scenario centres on the idea that very little is going to change over the next decades. It assumes that the world's leaders will take a blinkered and isolationist approach, protecting existing sources of energy and their economies at the expense of peace and freedom. Wars will be fought over oil because that will be the only way to secure supplies of a vital, dwindling resource. Part of the "solution" will be to keep on drilling, pushing the boundaries of exploration into previously sacrosanct wildernesses and areas previously thought to be inaccessible. Oil will be cleaned and refined in unconventional fields, wherever it can be found, and we will continue to pump millions of tonnes of carbon dioxide into the atmosphere. The effects of climate change, which will be accelerated by our continued and growing use of oil, will create more fear and encourage isolationism. Trade agreements will falter, crop yields will fall, and the world will experience rising unemployment, recession, and civil unrest. We will be supine in allowing oil reserves to be depleted, the environment to be degraded and international political co-operation to break down.

The "progressive solutions" scenario

This scenario sees the world embracing solutions that are already, or almost, at hand - within our technological grasp - and the development of a shared international vision: that the world is quite capable of thriving on the finite resources which it has, and that it can successfully make the transition from a hydrocarbon energy model to a mixed model. Natural gas will be used as a bridge fuel to a non-fossil fuel future, or one in which such fuels are used increasingly sparingly, while investment is made in clean-coal technologies. The use of renewables like wind and solar power will be maximized and while these will provide less energy than we have access to right now, a committed effort to conserve energy would enable us to generate the same degree of wealth using far less energy. We will begin to significantly reduce our greenhouse gas emissions and slow and perhaps finally even halt climate change, ensuring that those areas of the world which will bear the brunt of higher temperatures and a raised sea level are given assistance as part of an international agreement. For this scenario to become reality, the world would have to turn its back on conflict and enter a new era of co-operation, in which it is understood that there is only one future for everyone.

Recommended Reading

Ali, Tariq, *Bush in Babylon: The Recolonisation of Iraq*
Al Rasheed, Madawi, *The History of Saudi Arabia*
Clarke, Duncan, *The Battle for Barrels: Peak oil myths & world oil futures*
Deffeyes, Kenneth S., *Beyond Oil: The view from Hubbert's Peak*
Goodstein, David, *Out of Gas: The End of the Age of Oil*
Heinberg, Richard, *The Party's Over: Oil, war and the fate of industrial societies*
Ismael, Jacqueline S., *Kuwait: Dependency and Class in a Rentier State*
Ismael, Tareq Y. and Jacqueline S., *The Gulf War and the New World Order: International relations of the Middle East*
Keay, John, *Sowing the Wind: The Mismanagement of the Middle East 1900–1960*
Klare, Michael, *Blood and Oil: How America's thirst for petrol is killing us*
Kleveman, Lutz, *The New Great Game: Blood and oil in central Asia*
Leggett, Jeremy, *Half Gone: Oil, gas, hot air and the global energy crisis*

Lewis, Bernard, *The Middle East: A brief history of the last 2,000 years*

Mobbs, Paul, *Energy Beyond Oil*

Phillips, Kevin, *American Theocracy: The peril and politics of radical religion, oil, and borrowed money in the 21st century*

Roberts, Paul, *The End of Oil: On the edge of a perilous new world*

Ruppert, Michael C., *Crossing the Rubicon: The decline of the American Empire and the end of the age of oil*

Simmons, Matthew R., *Twilight in the Desert: The coming Saudi oil shock and the world economy*

Tertzakian, Peter, *A Thousand Barrels a Second: The coming oil break point and the challenges facing an energy dependent world*

Unger, Craig, *House of Bush, House of Saud: The secret relationship between the world's two most powerful dynasties*

Vidal, Gore, *Dreaming War: Blood for oil and the Cheney–Bush junta*

Yapp, M. E., *The Near East since the First World War: A history to 1995*

Yates, Douglas A., *The Rentier State in Africa: Oil rent dependency and neo-colonialism in the Republic of Gabon*